DEFINING A COMMON PLANNING FRAMEWORK FOR THE AIR FORCE

Leslie Lewis
Bruce Pirnie
William Williams
John Schrader

Prepared for the United States Air Force

RAND

PREFACE

This research was undertaken at the request of the U.S. Air Force's Strategic Planning Directorate. The task objective was to develop and help implement a common Air Force planning framework based on the strategies-to-tasks (STT) framework. Within the Air Force, the resourcing requirements and recommended allocations are developed within the Major Commands (MAJCOMs), and the Headquarters Air Force uses the Corporate Structure and the Air Force Board of Directors to look across all Air Force requirements and set institutional priorities. This approach was designed to identify and achieve the long-term institutional goals that provide strong linkages to important Department of Defense (DoD) resource allocation and management processes, in addition to complementing and enhancing existing Air Force planning and programming processes.

The purpose of this research was to develop a common planning framework that could extend across the Air Force, allow the corporate Air Force to develop and adjudicate requirements and resourcing options, and link to the external environment. The Directorate specifically asked that the framework incorporate the Air Force vision and capture nonoperational demands, and that the STT methodology provide the foundation for the framework. It wanted a structure and process designed that captured the totality of Air Force resources (e.g., operational, operational support, and institutional functions). A working group composed of representatives from each of the MAJCOMs and the functional organizations worked with RAND to generate design criteria. This report documents the development and subsequent rejection of the RAND-developed common planning framework. The study was conducted at RAND under the Strategy and Doctrine Program within Project AIR FORCE.

PROJECT AIR FORCE

Project AIR FORCE, a division of RAND, is the Air Force federally funded research and development center (FFRDC) for studies and analyses. It provides the Air Force with independent analyses of policy alternatives affecting the development, employment, combat readiness, and support of current and future air and space forces. Research is performed in three programs: Strategy and Doctrine, Force Modernization and Employment, and Resource Management and System Acquisition.

CONTENTS

Preface iii

Figures vii

Summary ix

Acknowledgments xv

Abbreviations xvii

Chapter One
- INTRODUCTION 1
- Problem 1
- Project Overview 5

Chapter Two
- DESIGN OF A PLANNING FRAMEWORK 7
- Design Criteria 7
- Additional Aims 8
  - Air Force Vision 8
  - Nonoperational Demands 9
- Approach 9
  - Choice of STT Methodology 9
  - Combatant Command Perspective 10
  - Service Perspective 10
  - Extension of STT Methodology 11

Chapter Three
- THE PLANNING FRAMEWORK 13
- Common Planning Framework 13
  - National Goals and Objectives 13

Joint Operational Objectives and Tasks 15
Supply in the Common Planning Framework 17
Definitions of Terms 17
Sources 19
Formulation 19
Planning Areas 20
Overview 20
Relationship to *Global Engagement* 21
Service Planning Areas 21
Foster High-Quality People 22
Evolve Through Innovation 23
Operational Planning Areas 24
Dominate Air and Space Operations 24
Attack Anywhere on the Globe 25
Provide Global Mobility 26
Achieve Global Awareness and Control of Forces 26
Provide Combat Support 27
Shape International Behavior 27

Chapter Four
IMPLEMENTATION OF A COMMON PLANNING FRAMEWORK 29

Appendix
A. MAJCOM DESCRIPTIONS 35
B. PLANNING AREAS 39

Bibliography 53

FIGURES

S.1. Common Planning Framework xii
3.1. Common Planning Framework, Highlighting Critical Demands 14
3.2. Common Planning Framework, Highlighting Planning Areas 18
3.3. Planning Areas 21
3.4. Global Engagement and Planning Areas 22
3.5. Service Planning Areas 23
3.6. Operational Planning Areas 25

SUMMARY

The Chief of Staff, Air Force (CSAF) concluded that the Air Force needed a mechanism to strengthen its corporate planning capabilities. The corporate planning function had to provide strong linkages to important Department of Defense (DoD) resource allocation and management processes, in addition to complementing and enhancing existing Air Force planning and programming processes. In 1997, a formal organization was established—the Air Force Strategic Planning Directorate (AF/XP)—with the purpose of linking Air Force planning and programming.

AF/XPX determined that the first step in improving corporate planning capabilities was to develop a common planning framework. A single planning framework could enable Air Force planners to identify corporate requirements; ensure their eventual resourcing; and justify Air Force near-, middle-, and long-term requirements to the external community—the Office of the Secretary of Defense (OSD), the Joint Staff (JS), and the other services.

CURRENT AIR FORCE PLANNING

The current Air Force planning mechanism is highly decentralized. Each Major Command (MAJCOM) uses a different planning framework. Although each of these is built upon a strategy-to-tasks (STT) concept, no single common planning framework exists within the Air Force. These individual frameworks are based upon missions for which the MAJCOMs provide Air Force capabilities.

Many of the MAJCOM planning frameworks are redundant, and they operate at varying levels of aggregation. Additionally, the individual frameworks focus primarily upon program planning, with an emphasis on modernization, often ignoring critical institutional functions. The MAJCOMs' frameworks mediate requirements and resourcing strategies within their assigned resourcing stovepipes, which are aligned primarily along core competencies. This stovepiped approach hinders the development of corporate Air Force options and affects total Air Force resourcing because there is no horizontal integration. Finally, the current process fails to link to the external environment, such as OSD, JS, and the other services because it is developed from an institutional rather than a joint perspective.

Creating a Common Planning Framework

Developing a common planning framework that could extend across the Air Force, allow the corporate Air Force to develop and adjudicate requirements and resourcing options, and link to the external environment required the creation of design criteria. The client asked that the framework incorporate the Air Force vision and capture nonoperational demands, and that the STT methodology provide the foundation for the framework. A working group composed of representatives from each of the MAJCOMs and the functional organizations worked with RAND to generate seven design criteria. According to the criteria, a common planning framework should

1. Display the elements that contribute to attaining a military capability
2. Be based on a hierarchy that links programs to national goals
3. Help identify intertemporal issues
4. Provide a basis for identifying and evaluating ways of attaining capabilities
5. Assist Air Force analysis and decisionmaking

6. Accommodate all data required for Air Force planning and programming[1]
7. Be understood by and be persuasive to all participants in the planning and programming processes, including the OSD, JS, unified commands, other services, and Congress.

The Planning Framework

The proposed common planning framework (Figure S.1) consists of eight planning areas (two service planning areas and six operational planning areas). The planning areas encompass demands that originate both inside and outside the Air Force. They are linked to the Air Force vision and the core competencies, the basic building blocks of the common planning framework. The shaded areas in the figure represent the two critical demands that the Air Force must meet: Air Force service functions and the missions of the unified commanders.

The proposed planning areas were derived primarily from *Global Engagement: A Vision for the 21st Century Air Force* (Department of the Air Force, 1997) and are designed to realize the vision of the Secretary of the Air Force (SAF) and the CSAF. This vision includes the core competencies, as well as other functions that extend across the Air Force. "Foster high-quality people" and "Evolve through innovation" are the two service planning areas that underlie and are intertwined with Air Force efforts in all operational planning areas.

The operational planning areas are related to the requirements of unified commanders and reflect the perspective that the Air Force is uniquely able to provide global reach and global power. The six operational planning areas are "Dominate air and space operations,"

[1]These data proceed from national military strategy, program guidance, Commander-in-Chief (CINC) requirements, acquisition programs, and PPBS (budgeting) inputs and outputs. In addition, there are unstructured data requests that concern revised fiscal guidance, modernization initiatives, changes in acquisition programs, and consideration of cost alternatives.

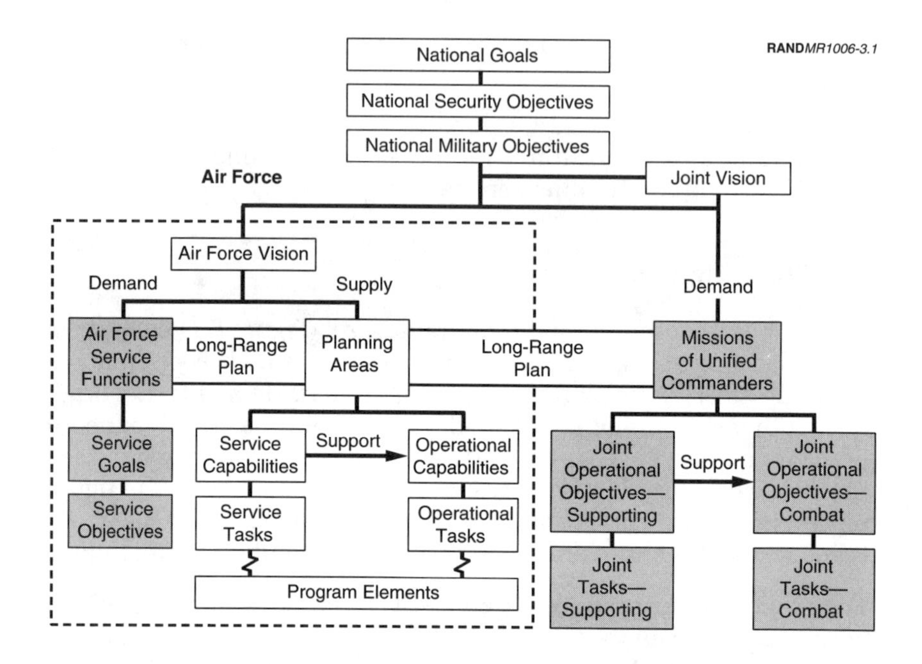

Figure S.1—Common Planning Framework

"Attack anywhere on the globe," "Provide global mobility," "Achieve global awareness and control of forces," "Provide combat support," and "Shape international behavior."

IMPLEMENTATION

Implementation is an integral part of the development and acceptance of the Air Force common planning framework. AF/XPX wanted the framework to be iteratively implemented over several planning periods. XPX leadership determined that the initial framework should be used in the development of the Air Force Long-Range Plan (LRP), which it was responsible for writing.

Prior to writing the LRP, XPX leadership shared the framework with its Board of Directors (BoD), which oversees and provides guidance to XPX on Air Force long-range planning issues. During the June 21, 1997, BoD meeting, the membership conceded the importance of using a common framework for planning and resourcing both operational and Title 10 responsibilities, but concluded that the proposed planning areas were confusing. The BoD determined that all Air Force planning and programming should be based on Air Force core competencies. This determination recreated the very problem that the common planning framework was attempting to solve: the predominance of the stovepipes that prohibited the Air Force leadership from identifying critical cross-cutting planning and resourcing issues. The Air Force leadership responded by creating thrust areas—issues that affect several core competencies for both planning and resourcing.

By early October 1997, the BoD was arguing that the thrust areas were not of sufficient depth to provide the necessary horizontal integration and should be abandoned. The Air Force leadership decided to terminate thrust areas at the January 20, 1999, Board of Directors meeting. The leadership has now decided to strengthen other existing processes, such as the programming panels and the BoD, to ensure that cross-cutting issues are raised and that the horizontal integration across MAJCOMs takes place. To attain a common integrating mechanism, the MAJCOMs have developed individual task lists that will link to the core competencies.

Although the Air Force chose not to implement the proposed common planning framework, the authors decided to document the effort nonetheless. The research raised some interesting issues and perspectives on planning, and we thought that this report would contribute to the literature and knowledge of defense planning and programming.

ACKNOWLEDGMENTS

The project team would like to thank Maj Gen David McIlvoy and Clark Murdock for their guidance and support throughout the project. We would also like to thank Lt Col Denise Ridgway for her efforts in creating the Working Group and for her collaboration on the planning framework. The following members of the Working Group provided invaluable assistance to the project team: Lt Col Bill Todd, Maj Rick Mayer, T. J. Sullivan, Kim Cain, Maj Ron Richard, Maj Delphine Rafferty, Maj Ronald Celentano, Daryl Fitzgerald, Capt Don Olds, Ken Lindsey, Bruce Seiber, William McKenna, Lt Col Chip Yarger, and Capt Tracy Nash.

At RAND we would like to thank Zalmay Khalilzad, Director of Project AIR FORCE's Strategy and Doctrine Program, for assisting the project team in its efforts. We would also like to thank Anissa Thompson for her research assistance and editorial support.

ABBREVIATIONS

AAW	Antiair warfare
ACC	Air Combat Command
AE	Aeromedical evacuation
AETC	Air Education and Training Command
AFMC	Air Force Materiel Command
AFRC	Air Force Reserve Command
AFSOC	Air Force Special Operations Command
AFSPC	Air Force Space Command
AFTL	Air Force Task List
AF/XP	Air Force Strategic Planning Directorate
AI	Air interdiction
AMC	Air Mobility Command
AMMP	Air Mobility Master Plan
APPG	Air Force Planning and Programming Guidance
ASW	Antisubmarine warfare
BM/C2	Battle Management and Command and Control
BoD	Board of Directors
CINC	Commander-in-Chief
CJCS	Chairman of the Joint Chiefs of Staff

CSAF	Chief of Staff of the Air Force
DCA	Defensive counterair
DoD	Department of Defense
DoDD	Department of Defense Directive
FYDP	Future years defense program
IO	Information operations
ISR	Intelligence, surveillance, and reconnaissance
IW	Information warfare
JROC	Joint Requirements Oversight Council
JS	Joint Staff
JTF	Joint Task Force
JWCA	Joint Warfighting Capability Assessment
LRP	Long-Range Plan
MAJCOM	Major Command
MAP	Mission Area Plan
MTW	Major theater war
NBC	Nuclear, biological, and chemical
NCA	National Command Authority
NMS	National Military Strategy
OSD	Office of the Secretary of Defense
PACAF	Pacific Air Forces
POM	Program Objective Memorandum
PPBS	Planning, Programming and Budgeting System
SAF	Secretary of the Air Force
SIOP	Single Integrated Operation Plan
STT	Strategy-to-tasks
TATPs	Thrust area transformation plans

UJTL	Universal Joint Task List
USAF	United States Air Force
USAFE	United States Air Forces in Europe
USSPACECOM	United States Space Command
USSTRATCOM	United States Strategic Command
WMD	Weapons of mass destruction

Chapter One

INTRODUCTION

In early 1995, the Chief of Staff of the Air Force (CSAF) determined that the Air Force needed to strengthen its corporate planning capabilities. The planning function had to link strongly to critical Department of Defense (DoD) resource allocation and management processes, such as the Planning, Programming, and Budgeting System (PPBS); Joint Warfighting Capability Assessment (JWCA); and the Joint Requirements Oversight Council (JROC). In addition, Air Force planning had to complement and enhance existing Air Force planning and programming processes. To meet these goals, the CSAF established a special assistant for long-range planning; in 1997, a formal organization was established (AF/XP) to link Air Force planning and programming.

In late 1996, the AF/XP determined that having a single corporate strategic planning process required developing a common planning framework. A single framework could enable Air Force planners to identify corporate Air Force requirements and ensure their eventual resourcing. A framework, if developed and implemented, could enable the Air Force to articulate and justify its near-, middle-, and long-term requirements to the external community—the Office of the Secretary of Defense (OSD), the Joint Staff (JS), and the other services.

PROBLEM

Historically, the planning function has been centralized in the Deputy Chief of Staff for Plans and Operations. The dilemma that the Air Force faces in defining and implementing a common planning framework is that current Air Force planning is a highly decentralized process that is conducted within the individual major commands (MAJCOMs) away from the Air Staff.[1] The common thread among the current processes is that they are based on a strategy-to-tasks (STT) concept, in which all tasks and their associated capabilities are derived from the national security strategy and support commander-in-chief (CINC) missions.[2] Although all the MAJCOMs use a STT framework, no single, common framework exists throughout the Air Force. The current planning process is built around each MAJCOM developing its own STT framework based on those missions for which it provides Air Force capabilities. The sets of capabilities are aligned under a list of Air Force–identified core competencies.[3] For instance, Air Combat Command (ACC) oversees all requirements and proposed resourcing for tasks associated with the Air and Space Superiority core competency. Air Superiority primarily supports the major theater war (MTW) mission.

As a consequence of decentralization, the number of parallel planning documents grew, fueled by MAJCOMs acting in isolation. Where responsibility for a program area was not clear-cut or the function cut across multiple areas, the MAJCOMs developed their own Mission Area Plan (MAP). Although there was only one plan for what were perceived as core MAJCOM missions, the number of supporting plans grew to (in some cases) one per command. As a consequence, there was no vantage point from which to make Air Force–wide strategic decisions. This issue surfaced as a major concern at the first MAJCOM working group meeting hosted by RAND. Planners felt that important initiatives were being lost in the process if the option cut across functional areas. Early in the project, a survey of planning documents produced several that treated Information Operations as a separate MAJCOM activity. None of these documents was coordinated outside the authoring command, although the Air

[1]The Air Force has nine MAJCOMs, which are responsible for conducting all phases of various missions and functions. Appendix A describes the MAJCOMs' missions.

[2]The STT framework, developed by RAND, defines objectives in a comprehensive hierarchy extending from fundamental national goals to tasks performed in combat.

[3]The Air Force's core competencies are Air and Space Superiority, Global Attack, Precision Engagement, Rapid Global Mobility, Information Superiority, and Agile Combat Support.

Force Information Warfare Center provided some informal integration of those plans it reviewed.

Many of the MAJCOM frameworks are redundant and operate at different levels of aggregation. For instance, Air Education and Training Command's (AETC's) activities are focused at a much lower level than those of ACC. AETC oversees all nonoperational training, some operational training, and education; thus, it is interested in classroom time and career training. On the other hand, ACC is responsible for the higher-level, operational development of future fighters to meet future mission requirements. To further underscore the dilemma, each MAJCOM has developed command-unique processes to support its own STT processes. ACC and Air Mobility Command (AMC) have mature processes with many levels of analyses and databases; AETC and Air Force Space Command (AFSPC) have less-developed processes and few, if any, supporting analytical capabilities.

The current STT process focuses primarily on program planning, with an emphasis on modernization. Often, critical institutional functions, such as nonoperational training, leader development, and quality-of-life issues, are not given full visibility in the resourcing process. Because the MAJCOMs are working from command-unique STT processes, there is little or no consistency in the terminology. A mission area to one MAJCOM might be a function to another.

The individual MAJCOM's requirements and resourcing strategies are adjudicated within their assigned resourcing stovepipe, which is aligned primarily along the core competencies. The Board of Directors (BoD) of AF/XP decided that the core competencies were too broad to establish meaningful priorities (Murdock, 1998, p. 8). Although the Air Force's core competencies are useful in shaping its vision and planning initiatives, they cannot be used independently to determine resource requirements. For example, core competencies do not contain sufficient details to link programming and resourcing, nor do they provide insights into how well you are doing in competing for resources across DoD.

The stovepiped approach hinders the development of corporate Air Force options and affects total Air Force resourcing because there is little horizontal integration. The most powerful MAJCOMs drive the resourcing priorities. Because the resourcing requirements and recommended allocations are developed within the MAJCOMs, the corporate Air Force has few mechanisms for viewing all Air Force requirements and setting institutional priorities. The stovepiped approach also inhibits the identification and achievement of long-term institutional goals. The inability to make strategic trades across functional areas when these functional areas became identified with a specific MAJCOM was recognized by some of the MAJCOMs.

Horizontal integration was possible only after the matter became a part of a MAJCOM MAP. What was needed was a single MAP or some other planning document in which the trades could be made prior to formalization in a command MAP. Integration of MAJCOM plans needed to happen at a much lower level to ensure that trade-offs were made. For example, the planning documents in Air Force Space Command could address migrating a system for aerospace command and control (e.g. AWACS) to space-based platforms, but did not address the more probable outcome from an operational perspective of an integrated space and air system that would rely on ACC platforms for contingency coverage. As a result, the MAJCOM plans were focused only on solutions well within the command's stovepipe functional areas. Also, space applications were slow to be included on ACC aircraft. This may have been a consequence of not having the means to look across the Air Force for new capabilities, only across a single command. AF/XP felt it needed the means to identify areas whereby cross-cutting innovative solutions and strategic trade-offs could be made before entering the more structured Air Force programming process.

The other problem, which the team tried to address for the Air Force, was the need to go beyond the Air Force for the objective demand. In the original project description, the planning framework was to be linked to joint processes. The team discovered that there also seemed to be institutional demand as well, so we broke demand into two sources. One collected external demands and the other collected internal demands. The Air Force had made little progress in documenting the necessity to resource in a manner that met long-term goals as well as the short-term immediate needs of the CINCs. The in-place system of MAPs and quality planning tried to place the need within major force programs that were essentially unchanged from pre-Cold War structures. Again, the emphasis was on programming and not planning.

PROJECT OVERVIEW

AF/XPX requested RAND's assistance in the development and implementation of a common Air Force STT framework that would complement and enhance existing Air Force planning and programming processes and would better link the Air Force vision to planning and programming. The client specifically wanted a structure and process that captured the totality of Air Force resources (e.g., operational, operational support, and institutional functions).

The proposed project contained four tasks to support the development and implementation objective of a common planning framework:

1. Identify and assess current Air Force STT efforts. The client wanted to apply the positive elements of existing processes to the corporate structure.
2. Assist XPX in the organization of a working group that would help develop and implement the framework. The client requested that the working group's membership be composed of operational and functional representatives from each of the MAJCOMs and the Air Force Staff.
3. Ensure that the proposed framework linked to the Air Force Long-Range Plan (LRP) and the Air Force Planning and Programming Guidance (APPG).[4]
4. Assist the Air Force in the implementation of the framework. XPX asked RAND to assist in assessing and proposing alternative implementation strategies to ensure that the framework would be iteratively implemented. The first phase would consist of gaining a consensus within the working group on the structure and at-

---

[4]The Air Force Strategic Plan (AFSP) replaced the LRP. The AFSP will extend out to three Future Years Defense Programs (FYDPs), or 17 years. The APPG spells out how the Air Force's various planning initiatives will be resourced in its Program Objective Memorandum (POM) and across the FYDP.

tributes of the framework. The working group members would then be responsible for assisting in the implementation within their respective organizations.

5. Work with XPX and the MAJCOM representatives to establish the ground rules for implementation of the framework across the Air Force.

The client wanted the framework developed and ready for implementation by late summer 1997. Although the client later decided not to implement this framework, the debate that it engendered has facilitated Air Force recognition of the need both to identify and to assess cross-functional issues.

Chapter Two

DESIGN OF A PLANNING FRAMEWORK

Developing a framework to extend across the Air Force first required developing the criteria. The framework also had to incorporate the current Air Force vision and explicitly capture nonoperational demands; these attributes were specified by the client. The client was also quite explicit that the STT methodology be the foundation of the framework, although attributes of the framework could be modified to accommodate the full spectrum of Air Force activities. AF/XPX leadership requested that institutional activities be captured, i.e., service-specific activities that act as broad enablers of operational capabilities.

DESIGN CRITERIA

Early in the project, the RAND-facilitated working group generated seven design criteria. According to these criteria, a common planning framework should

1. Display the elements that contribute to attaining a military capability
2. Be based on a hierarchy that links programs to national goals
3. Help identify intertemporal issues
4. Provide a basis for identifying and evaluating ways of attaining capabilities
5. Assist Air Force analysis and decisionmaking

6. Accommodate all data required for Air Force planning and programming[1]
7. Be understood by and be persuasive to all participants in the planning and programming processes, including the OSD, JS, unified commands, other services, and the Congress.

ADDITIONAL AIMS

Beyond satisfying these design criteria, the project had two additional aims: (1) incorporate as much as possible of the current Air Force vision and (2) explicitly capture nonoperational demands.

Air Force Vision

In recent years, the Air Force leadership has developed a vision of air power. The Secretary of the Air Force (SAF) and CSAF articulated this vision at the highest level in *Global Engagement: A Vision for the 21st Century Air Force* (Department of the Air Force, 1997). The vision was further developed through long-range planning initiatives that emerged from the fall 1996 CORONA and the spring 1997 CORONA.[2] To some extent, the vision is also reflected in MAJCOM mission areas and their supporting analyses. The vision provides the broad outline of a common planning framework and is widely known and understood within the Air Force. We incorporated the Air Force vision into the recommended framework to the greatest extent possible.

---

[1]These data proceed from national military strategy, program guidance, CINC requirements, acquisition programs, and PPBS inputs and outputs. In addition, there are unstructured data requests that concern revised fiscal guidance, modernization initiatives, changes in acquisition programs, and consideration of cost alternatives.

[2]The fall 1996 CORONA identified core competencies and 46 planning initiatives that were later collapsed into five operational thrusts: (1) the ability to find, fix, track, target, and engage anything of significance located or moving on, above, or below the surface of the earth; (2) expeditionary forces to provide tailored full-spectrum forces capable of rapidly deploying and delivering decisive air and space power on demand anywhere; (3) global command and control to support decisionmaking and decisive execution, at any level from local to global; (4) seamless control of the air and space environment and of the supporting surface environment and information infrastructure to secure protection from attack and freedom to operate; and (5) enabling capabilities that provide the essential underpinning for a military service. The spring 1997 CORONA refined these initiatives.

Nonoperational Demands

Early in the project, we recognized the importance of including nonoperational demands in the framework. Nonoperational demands are generated within the Air Force to meet standards that the Air Force sets for itself, such as quality of life for Air Force personnel and their families. Of course, nonoperational demands are ultimately linked to operational demands made by the National Command Authority (NCA) and the unified commanders. For example, the Air Force maintains a high quality of life to attract and retain the personnel necessary to ensure operational success. Nevertheless, nonoperational demands have their own dynamics and must be explicitly included in any comprehensive planning framework.

APPROACH

Although RAND agreed that the STT methodology was an ideal choice for a planning framework, the methodology had to be extended to accommodate different perspectives between combatant commands and services. Both ultimately work to a common end, but they have different perspectives that must be reflected in a planning framework.

Choice of STT Methodology

At the inception of the project, the client specified that the common planning framework employ STT methodology. STT is an ideal choice because its practical, commonsense approach is well understood within the Air Force. Moreover, it is intuitively persuasive to people outside the Air Force. It links lower-level objectives to national strategy in a clearly defined hierarchy, thus generating rationale and justification for programmatic decisions. It allows decisionmakers to review the strategic and operational effects of their decisions in an orderly, comprehensive way. However, the STT methodology was originally intended to support development of operational concepts, not the full range of functions performed by armed services.

Combatant Command Perspective

In its original form, STT reflected largely the perspective of combatant commands.[3] The Air Force focus is on providing warfighting capabilities based on operational concepts. Essentially, it defines objectives at each level of war from the nation's historical goals to immediate tactical aims. It reflects a classic hierarchy of strategy, operational art, and tactics that links the President to unit commanders and even individual pilots and others involved in the war. At every level, it asks what the commander wants to achieve.

STT was originally developed to support the development of operational concepts that employ weapons and techniques to produce desired combat outcomes. The basic thrust was operational and tactical, centering on the objectives of theater commanders, component commanders, and unit commanders engaged in combat. The methodology challenged developers to envision concepts, often implying new weapons and emerging technologies, that would enable commanders to accomplish their objectives more rapidly, more effectively, more surely, or at less risk to friendly forces.

Service Perspective

As a service, the Air Force is charged under U.S. Code Title 10 to perform many broadly defined functions that include every aspect of military forces except their actual employment in war, which is the responsibility of combatant commanders.[4] In a formal sense, the services provide forces to combatant commands, but they do more than that. They provide forces so sized, equipped, and trained that they can attain objectives of critical importance. In short, they provide capabilities. The Air Force does not provide forces to combatant

---

[3]STT has a long intellectual history at RAND, beginning with Glenn Kent's initiative in the early 1980s. See, among others, Kent and Simmons (1983), Lewis (1983), and Pirnie (1996).

[4]The Title 10 functions are (1) recruiting, (2) organizing, (3) supplying, (4) equipping, (5) training, (6) servicing, (7) mobilizing, (8) demobilizing, (9) administering, (10) maintaining, (11) constructing, outfitting, and repair of military equipment and (12) constructing, maintaining, and repair of buildings, structures, and utilities and the acquisition of real property and interests in real property necessary to carry out the responsibilities specified. (Public Law 99-433, October 1, 1986.)

commanders without suggesting how they should be employed. It prepares forces to make the greatest possible contributions to attaining CINC operational objectives. The Air Force defines and assesses various operational concepts to determine the capabilities that air power can provide. It expects to accomplish most, if not all, of its operational objectives in a joint context that implies both mutual support and competition among the services. Its operationally oriented programs link easily to the operational objectives of combatant commanders, although not always in one-to-one relationships. For example, a program to develop a "multicapable" aircraft may be linked to several operational objectives, including air superiority, degradation of an opponent's warmaking potential, and domination of land operations. But programs that are not operationally oriented may not link so directly to any operational objectives or may link indirectly to a range of objectives. For example, higher education doubtless helps develop Air Force officers into better warfighters and more effective representatives of their service but does not link directly to any particular operational objective. These types of activities are what we term *institutional.*

Extension of STT Methodology

To accommodate the totality of Air Force activities (i.e., operational and institutional activities), STT methodology must be extended to include service objectives that are not directly linked to a particular operational objective but that contribute generally to accomplishing several or all of them. These service objectives may be understood as broad enablers analogous to noncombat operational objectives. Just as deployment and sustainment underlie accomplishment of the entire range of combat objectives, higher education helps to produce Air Force officers who are more effective in accomplishing practically any objective.

In extending the STT methodology, it is important to keep the priority of combat objectives in mind. Like all armed services, the Air Force exists ultimately to accomplish objectives that will allow the NCA to impose its will on an opponent. These objectives have priority and determine other objectives. Extending the STT methodology horizontally does not imply that objectives residing on the same level have the same priority. On the contrary, combat objectives have priority; other objectives—whether noncombat objectives of theater commanders (such as deployment of forces) or service objectives (such as higher education for officers)—are important because they contribute to accomplishing combat objectives.

Chapter Three

THE PLANNING FRAMEWORK

This chapter describes a proposed set of planning areas that encompasses demands originating from both inside and outside the Air Force. From within, the Air Force strives to maintain the high quality of its men and women and to keep pace with technological advances. Success in these areas ensures that the Air Force can satisfy external demands, including the missions of unified commanders and direct tasking from the NCA. These planning areas are linked to the Air Force vision and core competencies, both of which are the basic building blocks of the common framework.

COMMON PLANNING FRAMEWORK

Figure 3.1 illustrates the common planning framework. The shaded areas depict the two critical demands that the Air Force must meet: Air Force service functions and the missions of the unified commanders. We will give an overview of the framework. More detailed discussion of the framework and its various attributes may be found in Appendix B.

National Goals and Objectives

The hierarchy begins with national goals and strategic direction from the NCA and culminates in the tasks necessary to accomplish those NCA goals and objectives. At the top of the common planning framework are the national goals—memorable statements of enduring national purpose found in historic documents (such as the

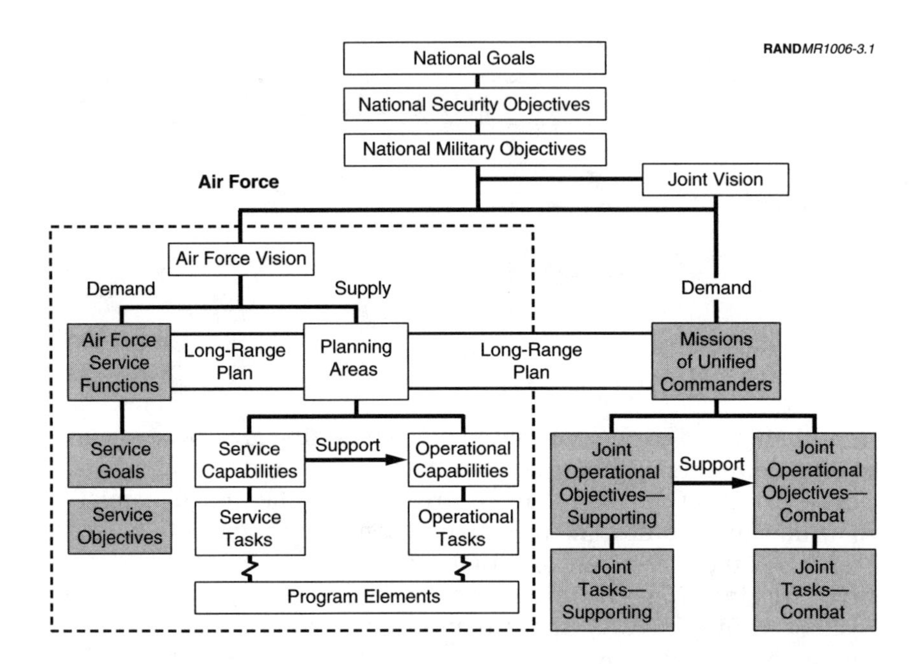

Figure 3.1—Common Planning Framework, Highlighting Critical Demands

Constitution, the Declaration of Independence, and the Gettysburg Address) that are set forth by statesmen. Public policy is aimed at achieving these goals.

National security objectives are derived from the President's National Security Strategy and include political, economic, and military means of protecting and defending fundamental U.S. interests and enduring goals (see Clinton, 1997). Unlike national goals, national security objectives are fluid and shift in accordance with changes in the geopolitical environment. "Enlarging the community of democratic, free-market countries" is an example of a national security objective.

The Chairman of the Joint Chiefs of Staff (CJCS) sets forth the national military objectives in the National Military Strategy (NMS) (Shalikashvili, 1997). National military objectives define the actions necessary to protect and defend U.S. principles, goals, and interests with respect to national goals and national security objectives.

"Promoting overseas presence," "maintaining peacetime engagement activities," and "responding to a full spectrum of crises" are some national military objectives.

National military objectives are attained by the missions and objectives of the unified commanders, supported by service capabilities. *Joint Vision 2010* (Shalikashvili, 1995) links the first three tiers of the framework with the internal Air Force side and the external joint side. The external demands are the missions of the unified commanders, which are in turn supported by

- supporting joint operational objectives,
- supporting joint operational tasks,
- combat joint operational objectives, and
- combat joint operational tasks.

The supporting objectives and tasks help the CINCs, component commanders, and Joint Task Force (JTF) commanders achieve combat objectives and tasks.

Joint Operational Objectives and Tasks

The external demands on the Air Force are imposed by the capabilities the unified commanders require to fulfill their missions. The missions of the unified commanders are defined in Joint Strategic Capability Plans and NCA orders. These missions state the intent of the NCA in broad political-military terms (for example, "deterring and defeating aggression against U.S. allies and friends") and are usually communicated through CJCS and the joint planning system.

Supporting joint operational objectives call for the provision of means or the creation of advantageous conditions to achieve the unified commander's mission. The sources for these objectives include campaign plans, the JWCA, *Universal Joint Task List* (UJTL) (CJCS, 1996),[1] the *Air Force Task List* (AFTL) (HQ USAF/XOOOE and

[1]One problem the services encounter in responding to extremely detailed task lists (such as the UJTL) is that the hierarchy of tasks is ambiguous; for example, "providing fuel" shares the same rank as "ensuring air superiority." RAND research supports the

AFDC), *Strategies-to-Tasks Baseline for USAF Planning,* and RAND documents that have tried to capture the generic elements of the joint commander's objectives (Kent and Simmons, 1991; Lewis et al., 1994; Lewis et al., 1995; Thaler and Shlapak, 1995; Pirnie, 1996). CINCs, component commanders, and JTF commanders determine the military force objectives at the operational level (for example, "gain information dominance"). In practice, each commander's site has specific campaign objectives derived from his or her tasking from the NCA.

Supporting joint tasks are objectives to be attained by military force at the tactical level. Unit and subunit commanders are responsible for accomplishing these tasks. At this level, the tasks outline the actions necessary to achieve the supporting joint operational objectives. For instance, acquiring intelligence on opposing forces and disrupting and distorting an opponent's information and intelligence would achieve the operational objective of "gaining information dominance." The supporting joint tasks are extracted from the JWCA, the UJTL, and RAND documents.

CINCs, component commanders, and JTF commanders are responsible for achieving combat joint operational objectives. These objectives are accomplished by the use of military force at the operational level. Combat joint operational objectives, such as "countering regional and global threats involving weapons of mass destruction (WMD)," are derived from campaign plans, RAND documents, and the UJTL.

"Assuring U.S. ability to operate in a WMD environment," "defending against attacks using WMD," and "suppressing and destroying opposing WMD" are examples of combat joint tasks culled from analysis, RAND documents, and the UJTL. Combat joint tasks consist of objectives that unit and subunit commanders and leaders are to attain through the use of military force at the tactical level.

The demands the unified commanders' missions place upon the Air Force must be balanced with internal Air Force demands. This requires addressing not just fiscal restraint but also the limits placed on

---

establishment of more general STT missions, objectives, and tasks that are based on CINC mission taskings.

the service by law and/or Department of Defense Directive (DoDD). These sets of demands must then be integrated to allow corporate Air Force leadership to make trade-offs and more capably represent the Air Force in the joint arena.

An example of an internal demand that would cause additional resources to be allocated in the programming cycle and finally in the service budget was the institutional need to limit the total number of days an Air Force member spent overseas in support of a joint force commander. The CSAF during our study stated that 120 days would be used as a standard. The consequence of this standard was to establish an internal demand, which increased the number of personnel needed to maintain the current CINC demand for some weapons systems. The old system would not recognize a demand such as this until, as in this case, a situation created a crisis in the ability of the service to maintain its commitment in the short term.

Supply in the Common Planning Framework

Figure 3.2 displays the common planning framework, with shaded areas depicting planning areas. The LRP embraces both the service functions of the Air Force and the missions of unified commanders.

Definitions of Terms

The following definitions were developed by RAND and refined during the working group meetings:[2]

- A **planning area** concerns Air Force senior leaders as they make decisions to ensure that the Air Force remains a successful military service and provides the best possible capabilities to unified commanders and to civil authorities. The proposed planning areas were derived primarily from *Global Engagement,*

[2]HQ AF/XP and RAND gathered a working group of experts in operations and various functional areas drawn from staffs in the MAJCOMs. Working group members met periodically at RAND's Washington Office and corresponded by electronic mail. In addition, AF/XP and RAND jointly visited the MAJCOM headquarters.

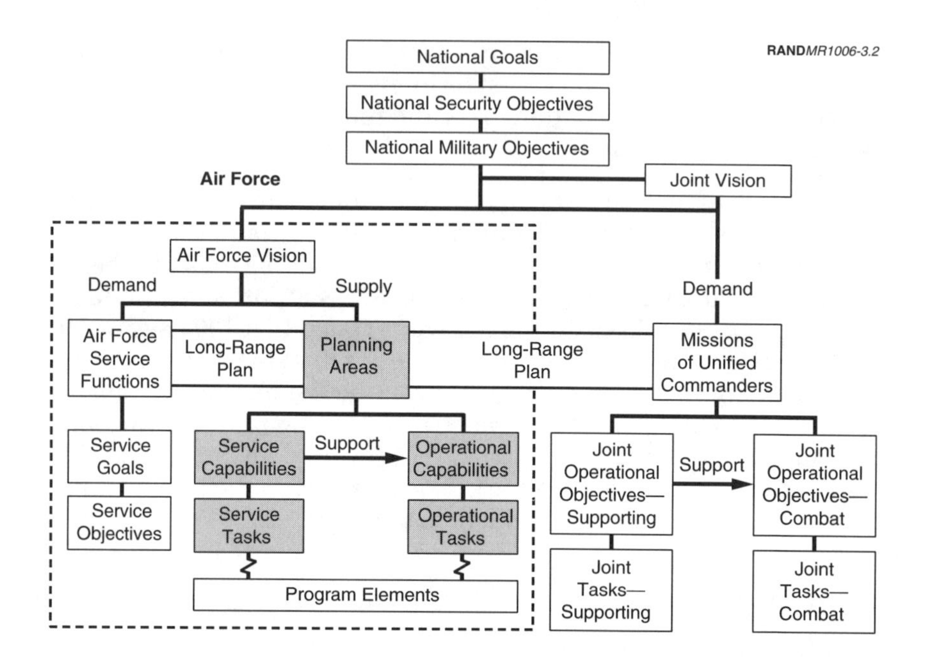

Figure 3.2—Common Planning Framework, Highlighting Planning Areas

especially the core competencies. The areas provide a means of assessing performance and raising issues to be addressed by Air Force senior leaders.[3]

- A **capability** is, broadly stated, the ability to progress toward a fundamental goal of the Air Force as a military service or to contribute to the attainment of the operational objectives of unified commanders or civil authority. Assessing capabilities helps to integrate related efforts at the levels of Air Staff, MAJCOMs, centers, and agencies. Through integrating the capabilities, planning area components can be determined.

---

[3]The expression "planning area" was chosen for its neutrality. The expression "core competency" was precluded because the proposed set of planning areas is not identical with core competencies, although it is closely related. The expression "mission area" was precluded because the proposed set is again not identical with mission areas, although there is much correspondence.

- A **task** is a more closely defined effort required to generate a capability.

Sources

The proposed set of planning areas are derived from several sources including documents and working group meetings. Air Force vision statements, especially *Global Engagement* (Department of the Air Force, 1997), are primary sources. The *Annual Report to the President and Congress* (Cohen, 1997) reflects concerns that cut across the military services. The UJTL provides an authoritative source of joint terminology. The AFTL contains helpful formulations. Mission Area Plans provide a wealth of detail at capability and task levels.

The Air Staff and the MAJCOMs helped develop the common planning framework during working group meetings and individual visits to their headquarters. During these visits, MAJCOM staff officers briefed the project team on the planning processes and priorities of their commands. The participating commands provided comments, most in writing, regarding the framework, planning areas, capabilities, and tasks. We considered these comments carefully and reflected many of them in the proposed set of planning areas. However, RAND is solely responsible for the material contained in this chapter.

Formulation

Every level in the common planning framework directs planning toward some desired outcome. Thus, every level could correctly be described as an objective. As far as is practical, capabilities and tasks are formulated without reference to the medium or means. For example, the capabilities implied by global attack are not associated with the mediums of either air or space because both might be involved. Similarly, one task is formulated as "ensure access to space," not "launch and recover satellites" because rocket launch may not always be the preferred means.

PLANNING AREAS

Overview

Through an iterative process of research and coordination, representatives of Air Force XP and the working group developed a set of planning areas (see Figure 3.3).

The planning areas allow Air Force planners to identify and categorize corporate Air Force requirements to ensure their eventual resourcing. Planning areas also help remove the stovepipes that hinder Air Force planning efforts because of a lack of horizontal integration. Horizontal integration provides the Air Force leadership with the ability to make trade-offs and to view options and alternatives.

The two service planning areas are enablers; they provide the foundation for operational planning areas. The six operational planning areas reflect how the Air Force contributes to full-spectrum dominance—dominating the nation's potential opponents across the entire range of military operations. Appendix B describes the planning areas and their elements.

There will not always be a one-to-one relationship between a program element and a planning area, much less between capabilities and tasks. On the contrary, one program element will often contribute to more than one planning area and will usually contribute to attaining more than one capability or accomplishing more than one task. For example, a multicapable aircraft, such as the F-16, helps to dominate air operations and to attack a wide variety of targets. This one-to-many relationship correctly reflects the broad utility of the weapon system.

The common planning framework provides a mechanism for linking operational objectives (from STT) with resource decisions. However, the framework will work only if resource decisions are made with an understanding of how each decision will affect service capabilities. Attaining this understanding will require assessments of capabilities from both the joint and Air Force perspectives. Program elements are only indirectly related to operational capabilities. They are a necessary part of the programing process, but it is the assessment of

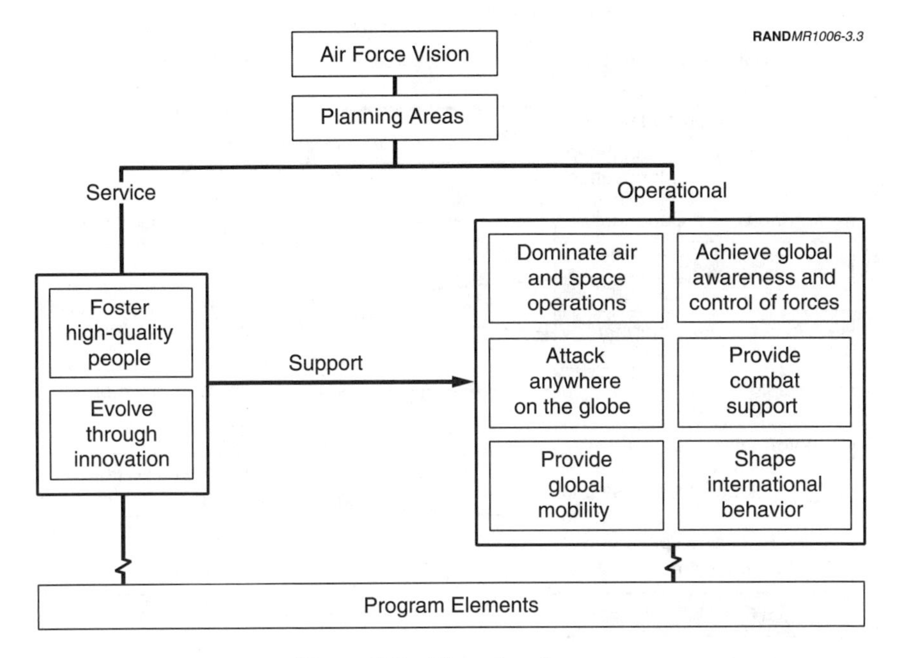

Figure 3.3—Planning Areas

the impact of a particular set of decisions on operational capabilities, involving many systems, that is relevant for senior decisionmaking.

Relationship to *Global Engagement*

The proposed set of planning areas is designed to realize and implement the vision of CSAF and SAF presented in *Global Engagement.* This vision includes the core competencies and other functions extending across the entire breadth of the Air Force. Figure 3.4 shows how the planning areas are derived from *Global Engagement.* Core competencies are enclosed in shaded boxes, while quotations from the text describe other planning areas.

Service Planning Areas

Service planning areas are broadly defined enablers, each supporting all the operational planning areas. For example, "maintaining a

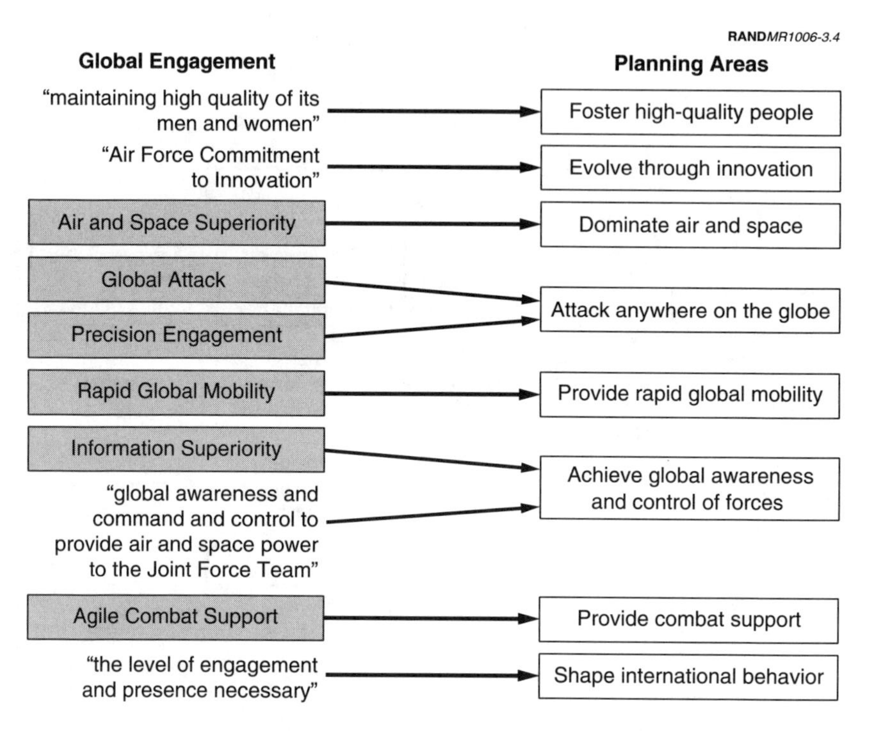

Figure 3.4—Global Engagement and Planning Areas

high-quality force" enables the Air Force to dominate air and space, provide rapid global mobility, and so on. These planning areas should give visibility to programs that do not directly serve the needs of unified commanders yet are fundamental to the Air Force contribution. The shaded area in Figure 3.5 shows the service planning areas.

Foster High-Quality People

Recruiting and sustaining the high quality of Air Force military and civilian personnel is a critical service issue. The need for high-quality personnel affects both the current Air Force and its future effectiveness. Success here means not only training people in technical skills but also imbuing them with the institutional vision and core values.

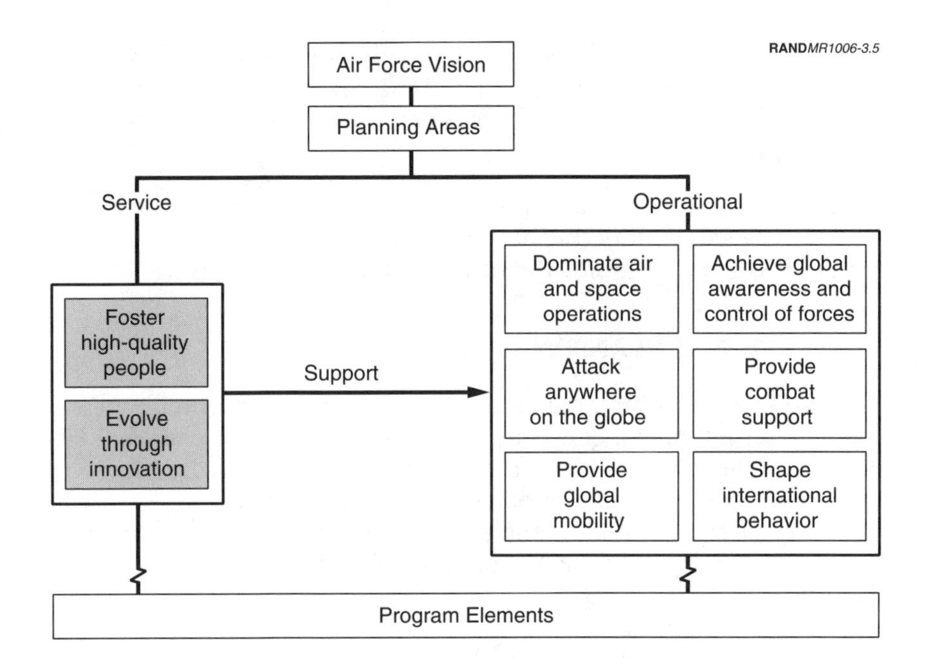

Figure 3.5—Service Planning Areas

In addition, the Air Force has an obligation to maintain a quality of life that fosters a wholesome community and family life. "People are at the heart of the Air Force's military capability, and people will continue to be the most important element of the Air Force's success in capitalizing on change." (Department of the Air Force, 1997, p. 19.)

The four elements of this planning area are accessing personnel, training and educating personnel, maintaining quality of life, and ensuring good order and fair treatment.

Evolve Through Innovation

The Air Force depends on technology and innovation to develop its operational capabilities. It is continually affected by technological change. Thus, innovation underlies every operational planning area. To maintain its edge over other nations, the Air Force must constantly develop new operational concepts and systems that effectively exploit emerging technology. "The key to ensuring today's Air Force core competencies will meet the challenge of tomorrow is *Innovation.*" (p. 10).

This planning area has four elements: providing analytic capability, sponsoring and conducting basic and applied research, empowering new ideas, and developing doctrine for air and space power.

OPERATIONAL PLANNING AREAS

Operational planning areas are related to the requirements of the unified commanders, generally on the level of joint operational objectives. These areas reflect the perspective of the Air Force as the service uniquely able to provide global reach and global power. The Air Force also generates air and space power to support civil authorities, for example, in disaster relief and interdiction of illegal drug traffic.

Some systems may help accomplish only one operational objective and therefore be considered only in that context. Other systems may help accomplish more than one operational objective and therefore be considered in several contexts. For example, variants of the multicapable F-16 may contribute to attaining three operational objectives (dominate air and space operations; attack anywhere on the globe; achieve global awareness). Such broad capability is an important rationale for developing multicapable systems and should be reflected in the planning framework. The shaded area in Figure 3.6 shows the operational planning areas.

Dominate Air and Space Operations

The Air Force has a unique ability to dominate operations in air and space, that is, to operate freely and deny freedom to opponents in the mediums of air and space. Sister services contribute to domination of air operations, but only the Air Force operates through the air-space continuum on a global basis. Success here is the fundamental, indispensable precondition for success in other planning areas.

The key elements in dominating air and space operations are suppressing air defenses; defeating air forces; suppressing and defend-

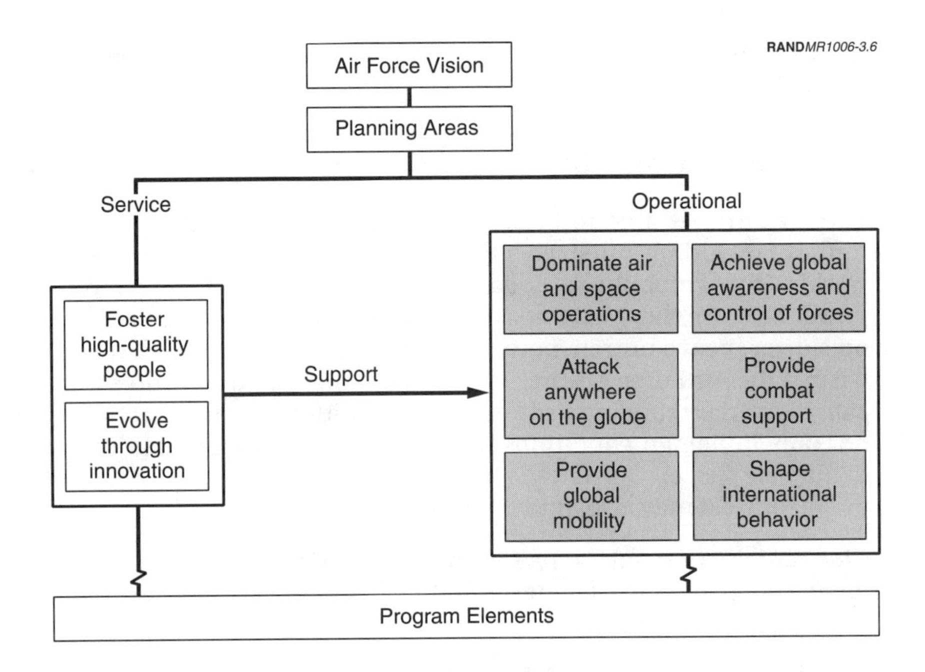

Figure 3.6—Operational Planning Areas

ing against cruise and ballistic missiles; and ensuring access to space, protecting friendly space assets, and countering opposing space assets.

Attack Anywhere on the Globe

The Air Force has a unique ability to attack targets located anywhere on the globe quickly and precisely. From its commanding position in air and space, the Air Force can use conventional weapons to engage any target that can be tracked. In concert with land and sea forces, the Air Force capability for global attack can dominate terrestrial operations. If necessary, the Air Force can employ nuclear weapons with the greatest selectivity possible for such devastating means.

The major elements of this planning area include projecting air forces globally, dominating land operations, dominating sea operations, degrading and destroying infrastructure, and countering weapons of mass destruction.

Provide Global Mobility

The Air Force is able to lift a wide range of passengers and cargo rapidly, including outsized military equipment, anywhere in the world. Global mobility is fundamental to America's status as a world power because there may not be enough warning to permit reliance on slower means of transportation. Even in the absence of suitable basing, the Air Force ensures global reach through the world's most capable aerial refueling fleet. It provides airlift and aerial refueling not only in benign conditions but also under combat conditions in denied air space. Currently, global mobility implies only airlift, but in the future it may also imply spacelift.

Global mobility requires the Air Force to provide lift in controlled or denied airspace, and to refuel aircraft in flight.

Achieve Global Awareness and Control of Forces

Immense leverage can be gained through information dominance and exploitation through timely and effective command and control.[4] The ability to view a battlespace in near real time—observing the movements of both friendly and opposing forces while blinding an enemy or distorting his vision—can give unified commanders crushing advantages. This planning area focuses attention on the revolution in military affairs made possible by microelectronics and computing: "Information technology advances will make dramatic changes in how this nation fights wars in the future."[5] The Air Force is a natural leader in this planning area because of its role in space and its extremely rapid, flexible applications of air and space power.

---

[4]These objectives are fundamental to military operations by all services at all levels of warfare from initial planning through execution and are therefore integral to other planning areas. However, the current pace of technological advance and the great expense of applying these technologies justify creation of a separate planning area.

[5]Ronald R. Fogleman and Sheila E. Widnall, "Cornerstones of Information Warfare," Department of the Air Force, 1997.

Achieving global awareness and control of forces involves providing worldwide communications and information systems; protecting worldwide communications and information systems; providing worldwide information; providing intelligence, surveillance, and reconnaissance; providing battle management and command and control; and attacking opposing information and control.

Provide Combat Support

The combat support planning area focuses on logistical support of air and space operations. It allows assessment of the entire logistical support system, from permanent installations in the continental United States to resupply of deployed forces. It includes oversight of logistic support associated with weapon systems and implementation of "best practices."

The goal is to improve operational support of the warfighting CINCs, including the efficiency of weapon system support, by pursuing "best value" processes and products. This calls for a full transition from deployed maintenance and "push" resupply to a method based on "accurate information, responsive production and daily, time-definite airlift."[6]

Achieving these goals for combat support involves supporting deployed forces, providing infrastructure to support air and space operations, and protecting forces.

Shape International Behavior

The Air Force contributes to the unified commands' shaping mission by interacting with foreign militaries and by equipping foreign forces. Forward presence, which may also shape behavior, is subsumed under the planning area "attack anywhere on the globe."

The shaping function has two principal elements—interaction with foreign militaries and equipping of foreign forces.

---

[6]Briefing, "CORONA Fall '96 Long-Range Planning, Summary Themes," slide 29.

Chapter Four

IMPLEMENTATION OF A COMMON PLANNING FRAMEWORK

Implementation is integral to the development and acceptance of a common Air Force planning framework. AF/XPX wanted the framework to be iteratively implemented over several planning periods, and concluded that the initial framework should be used in the development of the Air Force Long-Range Planning Guidance, which it was responsible for writing. The document would be read by all planners in the Air Force and would therefore provide a natural implementation mechanism.

AF/XPX first needed to share the proposed framework with its Board of Directors (BoD)[1] during its June 21, 1997 meeting. The BoD provides guidance on Air Force long-range planning issues. Before that meeting, an XPX action officer and members of the RAND project team discussed the framework and its various elements with the various MAJCOM planning groups and functional and action officers.

During that meeting, the BoD concluded that, while its members recognized the importance of a common framework for planning and

[1]The Air Force Planning BoD is chaired by the Vice Chief of Staff and consists of the Headquarters Air Force Assistant Secretaries, Deputy Chiefs of Staff, Vice Commanders from each Air Force MAJCOM, and advisers as needed. Its mission is to ensure development and implementation of the Air Force strategic and long-range planning effort. It advises the Air Force senior leadership on the status of planning efforts and ensures that planning and programming efforts reflect their priorities.

The BoD is also responsible for supporting the development of the Air Force's Strategic Vision and ensuring its implementation through the LRP, Strategic Plan, and APPG.

resourcing both the operational and Title 10 requirements, the proposed planning areas were confusing in light of the work done by the Air Force on the development of an agreed-upon set of core competencies. All Air Force planning and programming should be based on the core competencies.

The XPX representatives indicated that the core competencies were too broad and lacked specificity for planning and resourcing purposes; they further noted that the core competencies were embedded in the various planning areas. Nonetheless, the BoD concluded that, while the concept of a common planning framework is valid, the framework should be based on the Air Force's core competencies. They further indicated that the MAJCOMs should continue to perform their various planning and programming activities and that the work at XPX was to integrate the outputs of the functional organizations and MAJCOMs.

After the BoD meeting, XPX and RAND realigned the planning areas along the core competencies and key enablers. The planning areas were now:

- Air and space superiority
- Global attack and precision engagement
- Rapid global mobility
- Information superiority
- Agile combat support, quality people, global awareness, and Command and control
- Innovation.

The capabilities and tasks were then realigned according to each core competency and enabler.

The realignment of the framework along core competencies recreated the very problem that the common planning framework was attempting to solve: the stovepipes that inhibited the Air Force leadership's ability to identify critical cross-cutting planning and resourcing issues. In response to this insight, the Air Force leadership determined that a horizontal integration mechanism was needed to identify the cross-cutting issues. The horizontal integration mechanism was called *thrust areas*—areas calling for special Air Force emphasis. They identified corporately approved, long-range priorities that will shape the future Air Force and its core capabilities. Six thrust areas were identified, three operational and three institutional. (Note: The AF BoD decided to terminate the thrust areas because they were too broad, potentially duplicative, and resource intensive.)

The operational thrust areas were:

- Conduct seamless aerospace force
- Maintain a credible nuclear deterrent
- Find, fix, track, target, and engage.

The institutional areas were:

- Be an expeditionary aerospace force
- Develop the airman of the future
- Shape infrastructure for future aerospace force.

The XPX refined the thrust areas as a means of providing the critical horizontal integration across the core competencies and the various MAJCOM and functional planning and resourcing activities. Transformation plans were to set the performance targets for each thrust area and establish the means and paths to transition from where the Air Force is now to where it wants to be in the future (Murdock, 1998). Thrust area transformation plans (TATPs) would then describe integrated paths to the future. TATPs would have been created by integrating functional requirements and incorporating long-range plan end states and decision points. To achieve required end states and capability targets, divestiture and investment were intrinsic in the TATP development process. The TATPs would also inform integrated, phased, and prioritized investment/divestment resource streams, in addition to providing the foundation for future POMs driven by long-range priorities.

The process of aggregating Air Force requirements to a thrust area begins with the identification of independent capabilities to ensure compatible timing, which is then adjusted for technological risk. The outcome of the initial step is the aggregation of all related capabilities drawn from numerous sources. The next step is to determine the major investment and divestiture options. Following this step, the integration effort will be technologically, but not yet fiscally, constrained.

The TATPs would then be synthesized through a corporate process involving the Air Force Group[2] (Murdock, 1998). The outcome was envisioned to be prioritization and phasing of the thrust areas to achieve capability targets. The corporate integration process would have required several trade-offs to stay within the constraints. Trade-offs would have been made between thrust areas to create and ensure an affordable modernization program with temporal considerations. The synthesized TATPs were to be adjusted over time to reflect technological developments and competing priorities and should be reviewed when the Vision is revisited every four years.

The BoD, with the assistance of XPXP, has concluded that the role of Air Force planning is to provide strategic direction to the MAJCOMs. The existing processes, such as the CORONAs and the BoD meetings, enable the Air Force to provide horizontal integration across the institution. This process ensures that both institutional and operational issues are addressed at the MAJCOM and senior leadership levels.

The Air Force leadership continues to struggle with the issue of who really has the responsibility for corporate Air Force planning: Should the process be handled by the MAJCOMs with some overarching headquarters' guidance? Or should the process be highly centralized with headquarters establishing the broad institutional guidance for the MAJCOMs, which would in turn develop and implement their own initiatives in response to the guidance?

Recently, the Air Force leadership decided that XPP should provide broad guidelines for change, but that the MAJCOMs should retain their planning and programming. The integration mechanism will

---

[2]The Air Force Group consists of the Air Force panels and other headquarters offices and is chaired by the Deputy Director of XPP, Brigadier General Tom Goslin. The Group provides senior-level (colonel and civilian equivalent) resolution of resource allocation and other issues prior to Air Force Board review. It also develops the overall integrated Air Force program for submission to the Air Force Board.

be the AFTLs, which are derived from the UJTLs. Although each MAJCOM developed its own unique task list, the lists provide a mechanism for horizontal integration by ensuring that all planning and resourcing activities are linked to the UJTLs and the broadly articulated Air Force plans contained in the LRP.

The common planning framework described in this report was not adopted in part because it ran counter to the UJTL and the AFTL that was designed to fit within the UJTL structure. UJTL and AFTL appear to provide an already existing hierarchy of objectives developed according to a strategies-to-tasks approach. They are essentially lists of objectives that extend from the strategic to tactical level and they are authoritative, i.e., formally accepted by the Joint Staff and the Air Force. AFTL is structured around the Air Force core competencies and it includes those tasks required to train, organize, and equip aerospace forces, i.e., functions performed by a service under Title 10. Since UJTL and AFTL are already in place, the Air Force would risk causing confusion if it were to adopt a different set of objectives to support its long-range planning.

Could the Air Force adopt the UJTL and AFTL to support its own planning process? Ostensibly it could, but serious flaws in these lists make them unsuitable:

- UJTL confuses the levels of war and violates the principle of jointness. Specifically, it places operational objectives at the tactical level and assigns them to services.
- AFTL recognizes that UJTL has confused the level of war, but its cure is worse than the disease. It asserts: "However, since aerospace forces operate at all levels of war, the AFTL contains tasks that may occur at the strategic and operational levels of war, as well as the tactical level of war." This statement is a non sequitur that prevents levels of war from serving any useful purpose.
- AFTL mingles service functions (educate, train, and equip) with operational and tactical objectives. For example, Air Force Task (AFT) 1.1—provide counterair capabilities—includes AFT 1.1.2—educate and train counterair forces—and AFT 1.1.3—equip counterair forces. This constant mingling of service functions with operational objectives makes AFT useful for training

purposes—e.g., development of mission essential task lists—but unsuitable to support resources allocation decisions.

By early October 1997, the BoD was arguing that the thrust areas were not of sufficient depth to provide the necessary horizontal integration and should be abandoned. The Air Force leadership decided to terminate thrust areas at the January 20, 1999, Board of Directors meeting. It will strengthen other processes, such as the programming panels and the BoD, to ensure that cross-cutting issues are raised and that horizontal integration across MAJCOMs takes place. To attain a common integrating mechanism, the MAJCOMs have developed individual task lists that will link to the core competencies.

Although the Air Force chose not to implement the proposed common planning framework, we document the RAND effort nonetheless. The research raised some interesting issues and perspectives on planning and we thought that this report would contribute to the literature and knowledge of defense planning and programming.

Appendix A

MAJCOM DESCRIPTIONS

The Air Force's MAJCOMs are responsible for conducting all phases of various missions and functions. There are currently nine MAJ-COMs, with the following missions:[1]

- **Air Combat Command** (ACC)—to operate U.S. Air Force (USAF) bombers and USAF continental U.S.-based combat-coded fighter and attack aircraft and combat support-coded reconnaissance, rescue, battle management, and command and control aircraft; to organize, train, equip, and maintain combat-ready forces for rapid force deployment and employment to meet the challenges of peacetime air sovereignty and wartime defense; to provide air combat forces to America's warfighting commands; and to provide nuclear-capable forces for U.S. Strategic Command (USSTRATCOM).
- **Air Education and Training Command** (AETC)—to recruit, access, commission, train, and educate Air Force enlisted and officer personnel; to provide basic military training, initial and advanced technical training, flying training, and professional military and degree-granting professional education; and to conduct joint medical service, readiness, and Air Force security assistance training.

---

[1]The following material was compiled from the following sources: U.S. Force Fact Sheets published on the World Wide Web (available through http://www.af.mil/news/indexpages/fs_index.html) and from "USAF Almanac," *Air Force Magazine*, May 1997.

- **Air Force Materiel Command (AFMC)**—to manage the integrated research, development, test, acquisition, and sustainment of weapon systems; to produce and acquire advanced systems; to operate "superlabs," major product centers, logistics centers, and test centers; and to operate the USAF School of Aerospace Medicine and USAF Test Pilot School.
- **Air Force Space Command (AFSPC)**—to operate and test USAF intercontinental ballistic missile forces for USSTRATCOM; to operate missile warning radars, sensors, and satellites; to operate national space-launch facilities and operational boosters; to operate worldwide space surveillance radars and optical systems; to provide command and control for DoD satellites; and to provide ballistic missile warning to the North American Aerospace Defense Command and USCINCSPACE.
- **Air Force Special Operations Command (AFSOC)**—to serve as the Air Force component to U.S. Special Operations Command; to deploy specialized air power, delivering special operations combat power anywhere, anytime; and to provide Air Force special operations forces for worldwide deployment and assignment to regional unified commands to conduct unconventional warfare, direct action, special reconnaissance, counterterrorism, foreign internal defense, counterproliferation, civil affairs, humanitarian assistance, psychological operations, personnel recovery, and counternarcotics operations.
- **Air Mobility Command (AMC)**—to provide rapid, global airlift and aerial refueling for U.S. armed forces; to serve as USAF component of the U.S. Transportation Command; and to support wartime taskings by providing forces to theater commands.
- **Pacific Air Forces (PACAF)**—to plan, conduct, and coordinate offensive and defensive air operations in the Pacific and Asian theaters and to organize, train, equip, and maintain resources to conduct air operations.
- **U.S. Air Forces in Europe (USAFE)**—to plan, conduct, control, coordinate, and support air operations to achieve U.S. national and the North Atlantic Treaty Organization objectives based on taskings by the United States European Command CINC.

- **The Air Force Reserve Command** (AFRC)—to support the active duty force; serve in such missions as fighter, bomber, airlift, aerial refueling, rescue, special operations, aeromedical evacuation, aerial firefighting, weather reconnaissance, space operations, and airborne air control; provide support and disaster relief in the United States; and to support national counterdrug efforts.

Appendix B

PLANNING AREAS

SERVICE PLANNING AREAS

Foster High-Quality People

Recruiting and sustaining the high quality of Air Force military and civilian personnel is a critical service issue. The need for high-quality personnel affects both the current Air Force and its future effectiveness. Success here means not only training people in technical skills but imbuing them with institutional vision and core values. In addition, the Air Force has an obligation to maintain a quality of life that fosters a wholesome community and family life. "People are at the heart of the Air Force's military capability, and people will continue to be the most important element of the Air Force's success in capitalizing on change." (Department of the Air Force, 1997, p. 19.)

This planning area has four elements:

- **Access Personnel**—Acquire and develop basic airmen, new officers, and entry-level civilian workers:
  - — Attract high-quality people through Air Force recruiting efforts: "we will continue our aggressive recruiting efforts to assure we continue to attract high-caliber people into our Air Force."[1]
  - — Conduct basic military training, producing basic airmen.

---

[1]CSAF Goals for 1997," *Policy Letter Digest*, January 1997, p. 2.

— Provide precommissioning programs, such as the Air Force Academy and the Reserve Officer Training Corps, to produce new officers.

- **Train and Educate Personnel**—Provide continuing professional development throughout Air Force careers, reinforcing the Air Force's "core values in all aspects of its education and training . . . throughout a career." (p. 21.) Moreover, the "civilian career development program needs to be improved to create the same institutional commitment and responsibility in all our people—military and civilians."[2]
- **Maintain Quality of Life**—Ensure that life in the Air Force will be wholesome and attractive for all its members. The Air Force places and will continue to place a high priority on "providing quality of life for Air Force members and their families." (p. 19.) It is also committed to preserving "a 'sense of community' at its bases maintaining the Quality of Life standards while searching for new and more efficient ways of providing them."[3]
- **Assure Good Order and Fair Treatment of Personnel**—Administer and enforce military justice and provide programs to ensure equal opportunity and absence of harassment.

Evolve Through Innovation

The Air Force depends on technology and innovation to develop its operational capabilities. It is continually affected by technological change. Thus, innovation underlies all the operational planning areas. To maintain its edge over other nations, the Air Force must constantly develop new operational concepts and systems that effectively exploit emerging technology. "The key to ensuring today's Air Force core competencies will meet the challenge of tomorrow is *Innovation.*" (p. 10.)

This planning area also has four elements:

---

[2]Briefing, "CORONA Fall '96 Long-Range Planning, Summary Themes," slide 23.

[3]Briefing, "CORONA Fall '96 Long-Range Planning, Summary Themes," slide 25.

- **Provide Analytic Capability**—Invest in analytic activities, including models and simulations.
- **Sponsor and Conduct Basic and Applied Research**—Invest in research whose military application is promising but still undetermined, including laboratories. These capabilities explore the potential for affecting military operations, rather than simply responding to operational needs. Battle laboratories with operational focus would fall in other planning areas.
- **Empower New Ideas**—Experiment with new ways to make progress in the operational planning areas, including test centers. "The Air Force is committed to a vigorous program of experimenting, testing, exercising and evaluating new operational concepts and systems for air and space power." (p. 9.) This would include setting up Battle Labs "to provide institutional, operational focus for testing, evaluating, and prototyping new concepts for Air and Space combat in the 21st Century."[4]
- **Develop Doctrine for Air and Space Power**—Work to formulate and promulgate the current understanding of how the Air Force will contribute to accomplishing the operational objectives that unified commanders and civil authorities set. Development of doctrine is an explicit Title 10 responsibility.

OPERATIONAL PLANNING AREAS

Operational planning areas are related to the requirements of unified commanders, generally on the level of joint operational objectives. These areas reflect the perspective of the Air Force as the service uniquely able to provide global reach and global power. The Air Force also generates air and space power to support civil authorities, for example, in disaster relief and interdiction of illegal drug traffic.

---

[4]Briefing, "CORONA Fall '96 Long-Range Planning, Summary Themes," slide 5.

Dominate Air and Space Operations

The Air Force has a unique ability to dominate operations in air and space, that is, to operate freely and deny freedom to opponents in the mediums of air and space. Sister services contribute to domination of air operations, but only the Air Force operates through the air-space continuum on a global basis. Success here is the fundamental, indispensable precondition for success in other planning areas.

This planning area corresponds to the core competency of Air and Space Superiority: "The control of air and space is a critical enabler for a Joint force because it allows all U.S. forces freedom *from* attack and freedom *to* attack. With air and space superiority, the Joint Force can dominate enemy operations in all dimensions—land, sea, air, and space." (p. 10.) This is also spelled out as "Gain and maintain air superiority in theater of war." (CJCS, 1996, p. 2-56.)

The key elements in dominating air and space operations are as follows:

- **Suppress Air Defenses**—Coordinate, integrate, and synchronize attacks that "neutralize, destroy, or temporarily degrade surface-based enemy air defenses by destructive and/or disruptive means." (CJCS, 1996, p. 2-112.)
- **Defeat Air Forces**—If necessary, defeat, rather than merely countering, attacking, or neutralizing opposing air forces. Air base attack and air-to-air engagement are implied in this capability. It involves countering enemy air attacks in a theater of operations by intercepting, engaging, destroying, or neutralizing enemy formations in the air, "using all available air-, land-, or sea-based air defense capabilities of the joint force to achieve operational result," (CJCS, 1996, p. 2-131), including attacks on aircraft and missiles (CJCS, 1996, p. 2-112).
- **Suppress and Defend Against Cruise and Ballistic Missiles**—Defend against the growing theater and global threat posed by cruise and ballistic missiles, which is "one of the developments accelerating warfare along the air-space continuum." (Department of the Air Force, 1997, p. 10.) The Air Force currently

considers these threats separately, but "over time it will merge them into a common missile defense architecture."[5]

- **Ensure Access to Space, Protect Friendly Space Assets, and Counter Opposing Space Assets**—Provide the capabilities required to dominate operations in the medium of space. This means ensuring "access to space, freedom of operations within the space medium, and an ability to deny others the use of space, if required." (U.S. Space Command, 1997.)[6] Since U.S. Space Command already supports the warfighter through its missions of space control and space force application, "the further militarization of space will be driven by international events, national policy, threats moving through and from space, and threats to US space assets." (Department of the Air Force, 1997, p. 17.) The Air Force must be prepared to move further at NCA direction.

Attack Anywhere on the Globe

The Air Force has a unique ability to attack targets located anywhere on the globe quickly and precisely. From its commanding position in air and space, the Air Force can use conventional weapons to engage any target that can be tracked. In concert with land and sea forces, the Air Force capability for global attack can dominate terrestrial operations. If necessary, the Air Force can also employ nuclear weapons with the greatest selectivity possible for such devastating means.

This planning area corresponds to the core competencies of Global Attack and Precision Engagement. Global attack refers to the ability to attack rapidly anywhere on the globe As part of that ability, the "Air Force will sustain its efforts in the nuclear area and strengthen its response to the growing risk of proliferation." (Department of the Air Force, 1997, p. 11.) Precision engagement refers to the Air Force's "ability to apply selective force against specific targets and achieve

---

[5]Briefing, "CORONA Fall '96 Long-Range Planning, Summary Themes," slide 18.

[6]The document also offers alternative formulations: "dominating the space dimension of military operations" and "the emerging synergy of space superiority with land, sea, and air superiority."

discrete and discriminate effects." (Department of the Air Force, 1997, p. 13).

The following are key elements of this planning area:

- **Project Air Forces Globally**—Provide the services needed to generate global power, except air refueling (which is collected in the "provide global mobility" planning area). The following must be considered in this context:
  - Forward base forces: Costs are incurred in basing air forces permanently in foreign countries. "U.S. forces permanently stationed and rotationally or periodically deployed overseas serve a broad range of U.S. interests." (Cohen, 1997, p. 6).
  - Preposition stocks: Stocks are maintained in forward locations to support air operations during crisis and war.
  - Deploy expeditionary forces: Air and space power contribute to engagement and presence "by augmenting those forces that are permanently based overseas with temporary or rotational deployments and power projection missions." (Department of the Air Force, 1997, p. 7). "The Air Force will develop new ways of doing mobility, force deployment, protection, and sustainability in support of the expeditionary concept." (Department of the Air Force, 1997, p. 11).
- **Dominate Land Operations**—Use new sensing technologies and sensor-fuzed weapons to allow air power to deny movement by day and night, fixing maneuver forces in disadvantageous positions. The tasks distinguish between interdiction of maneuver forces and close air support with the implied requirement for precise and flexible forward air control, usually in cooperation with U.S. and allied land forces engaging an enemy.
- **Dominate Sea Operations**—Gain and maintain maritime superiority in a theater of war. This task attacks the enemy's warfighting capabilities at sea via antisubmarine warfare, antiair warfare, defensive counterair, air interdiction, and traditional surface and subsurface warfare. Antiair warfare and defensive counterair are encompassed in the planning area "dominate air and space." (CJCS, 1996, p. 2-56.) Maritime operations are a collateral mission for the Air Force under Title 10.

- **Degrade and Destroy Infrastructure**—Reach over the battle area and attack an enemy in full strategic depth, reducing its war making potential. The goal is to destroy or neutralize strategic-level targets and to shape and control the tempo of theater campaigns and joint operations, "using all available joint and allied firepower assets against land, air (including space), and maritime (surface and subsurface) targets having strategic significance." (CJCS, 1996, p. 2-62.)
- **Counter Weapons of Mass Destruction**—Improve the nation's ability to deter and prevent the effective use by an adversary of nuclear, biological, and chemical weapons, to defend against them, and to fight more effectively in an environment in which such weapons are used. (Cohen, p. 5.) In recognizing the dangers posed by the efforts of other nations to acquire nuclear weapons, the "Air Force will sustain its efforts in the nuclear area and strengthen its response to the growing risk of proliferation."[7] (Department of the Air Force, 1997, p. 11.) Weapons of mass destruction upset regional balances, promote intimidation, and force the "have nots" to seek them as well. "Therefore, we must be able to **find, track, and neutralize an adversary's WMD Capability.**"[8]

Provide Global Mobility

The Air Force is able to lift rapidly a wide range of passengers and cargo, including outsized military equipment, anywhere in the world. Global mobility is fundamental to America's status as a world power because there may not be enough warning to permit reliance on slower means of transportation. Even in the absence of suitable basing, the Air Force ensures global reach through the world's most capable aerial refueling fleet. It provides airlift and aerial refueling not only in benign conditions but also under combat conditions in

---

[7]Although nuclear weapons may pose the most serious risk, rogue states might also seek to develop biological and chemical weapons, as Iraq did prior to the Gulf War; therefore, we preferred the broader term "weapons of mass destruction."

[8]Special Operations Command, *SOF Vision 2020*, MacDill Air Force Base, Florida, 1997, p. 6.

denied air space. Currently, global mobility means airlift, but in the future it may include spacelift.

This planning area corresponds to the core competency of Rapid Global Mobility, providing the nation its global reach and underpinning its role as a global power: "When an operation must be carried out quickly, airlift and aerial refueling will be the key players." (USAF, 1997, p. 12). A primary source for capabilities and tasks is the *Air Mobility Master Plan* (AMMP) (AMC, 1997).

Global mobility requires the Air Force to

- **Provide Lift in Controlled Airspace**—Transport personnel and cargo through airspace controlled by friendly forces. This includes lifting
  - — Combat and support personnel, including "unit rotations, and movement of the President and senior government or executive personnel" (AMC, 1997, p. 1-12)
  - — Regular passengers, that is, military personnel and civilians without special requirements
  - — Very important persons, that is, military and civilian officials with rank or status requiring special amenities, security, and communications
  - — Medical patients (aeromedical evacuation, providing "rapid worldwide transportation of ill or injured personnel to appropriate medical care" (AMC, 1997, p. 1-10)
  - — Materiel and equipment, delivering "supplies and equipment whose urgency or nature cannot wait for surface transportation" (AMC, 1997, p. 1-10).
- **Provide Lift in Denied Airspace**—Transport personnel and cargo when opposing forces might seek to deny access. This includes
  - — Providing combat delivery (airdrop), unloading personnel or material from aircraft in flight. "This combat employment and resupply of forces is used when the airland option is not available." (AMC, 1997, p. 1-10.)
  - — Lifting special operations forces, providing specialized airdrop support to special operations forces for joint or com-

bined training, contingencies, operations other than war, and other missions. (AMC, 1997, p. 1-13.)

— Rescuing personnel and recovering equipment, rescuing downed air crews and sensitive items of equipment from denied areas. (CJCS, 1996, p. 2-134.)

- **Refuel Aircraft in Flight**—Using air tankers for refueling during
  — Normal operations, to allow "rapid deployment of fighters, bombers, and combat support aircraft" (AMC, 1997, p. 1-11.)
  — Special operations, to provide air refueling of joint, multinational, or special operations aircraft, as distinguished by the customer's unique requirements. Air refueling requires special equipment, specialized crew training, and modified operational procedures. Today's refueling fleet was originally developed to support strategic nuclear bombers. Such missions have additional associated hazards, because they "may be conducted in a nuclear detonation environment, leading to electromagnetic pulse, flash blindness, and routing problems." (AMC, 1997, p. 1-10.)

Achieve Global Awareness and Control of Forces

Immense leverage can be gained through information dominance and its exploitation by using timely and effective command and control.[9] The ability to view a battlespace in near real time—observing the movements of both friendly and opposing forces while blinding an enemy or distorting his vision—can give unified commanders crushing advantages. This planning area focuses attention on the revolution in military affairs made possible by microelectronics and computing: "Information technology advances will make dramatic changes in how this nation fights wars in the future."[10] The Air Force is a natural leader in this planning area because of its leading role in

[9]These objectives are fundamental to military operations by all services at all levels of warfare from initial planning through execution and are therefore integral to other planning areas. But the current pace of technological advance and the great expense of applying these technologies justify creation of a separate planning area.

[10]Ronald R. Fogleman and Sheila E. Widnall, "Cornerstones of Information Warfare," Department of the Air Force, 1997.

space and its extremely rapid, flexible applications of air and space power.

This planning area corresponds to information superiority in the *Joint Vision:* "Information superiority will require both offensive and defensive information warfare (IW). Offensive information warfare will degrade or exploit an adversary's collection or use of information." (Shalikashvili, 1995, p. 10.) "Gain and maintain information superiority in theater of war/area of responsibility" is defined as achieving

> information superiority by affecting an adversary's information, information-based processes, and information systems, while defending one's own information, information-based processes, and information systems. (CJCS, p. 2-56.)

According to former Air Force Chief of Staff General Ronald Fogleman and former Air Force Secretary Sheila Widnall, "Information Warfare has become central to the way nations fight wars, and will be critical to Air Force Operations in the 21st century."[11]

This planning area also corresponds to the core competency of Information Superiority:

> While Information Superiority is not the Air Force's sole domain, it is, and will remain, an Air Force core competency. The strategic perspective and the flexibility gained from operating in the air-space continuum make airmen uniquely suited for information operations. . . . Providing Full Spectrum Dominance requires a truly interactive common battlespace picture. (Department of the Air Force, 1997, p. 14.)

This includes Global Awareness and Command and Control: "These competencies are brought together by global awareness and command and control to provide air and space power to the Joint Force Team." (Department of the Air Force, 1997, p. 9.)

Achieving global awareness and control of forces involves

---

[11]Ibid.

- **Providing worldwide communications and information systems**—This is fundamental to all information warfare.
- **Protecting worldwide communications and information systems**—"The top IW priority is to defend our own increasingly information-intensive capabilities." (Department of the Air Force, 1997, p. 14.) Information assurance is defined as information operations that "protect and defend information and information systems by ensuring their availability, integrity, authentication, confidentiality, and non-repudiation. This includes providing for restoration of information systems by incorporating protection, detection, and reaction capabilities."[12]
- **Providing worldwide information**—In the near term, the highest payoff here is in providing worldwide intelligence, surveillance, and reconnaissance. (Department of the Air Force, 1997, p. 14.) The Air Force must provide "reliable and effective space-based navigation, communications, and weather support."[13]
- **Providing intelligence, surveillance, and reconnaissance**—The focus here is on theater strategic intelligence, surveillance, and reconnaissance. (CJCS, p. 2-57.)
- **Providing battle management and command and control**—The commitment is to provide an integrated picture of the global and theater air, space, and surface battlespace for the commander of the joint force of the 21st century. Future battle management and command and control (BM/C2) systems are to "enable real-time control and execution of all air and space missions." (Department of the Air Force, 1997, p. 14.) This has been applied in Bosnia, where structuring the air operations around the need to "get these information platforms overhead" has been "the core of the Air Tasking Order, driving all else."[14]

---

[12]Unclassified excerpt, DoD Directive S-3600.1, Information Operations, 3 December 1996.

[13]"1996 Stakeholder's Report," *Guardian*, February 1997, p. 4.

[14]Secretary of the Air Force Sheila E. Widnall, speech given at Goodfellow Air Force Base, Texas, June 18, 1997.

- **Attacking opposing awareness and control**—The focus here is on using information warfare offensively.[15] The emphasis is operational and tactical, and the Air Force will "continue, in conjunction with other Federal agencies, to support strategic information operations." (Department of the Air Force, 1997, p. 14.) The Air Force conducts normal and special information operations, which are "taken to affect adversary information and information systems while defending one's own information and information systems." Special information operations are sensitive because of "their potential effect or impact, security requirements, or risk to the national security of the U.S."; these require a special review and approval process.[16]

Provide Combat Support

The combat support planning area focuses on logistical support of air and space operations. It allows assessment of the entire logistical support system, from permanent installations in the continental United States to resupply of deployed forces. It includes oversight of logistic support associated with weapon systems and implementation of "best practices."

This planning area is related to the core competency of Agile Combat Support and to "Key Elements of Air Force Infrastructure" in *Global Engagement:* "Agile Combat Support is recognized as a core competency for its central role in enabling air and space power to contribute to the objectives of a Joint Force Commander." (Department of the Air Force, 1997, p. 16.) "All support activities will be run more like businesses, using the 'best practices' gleaned from top performers." (Department of the Air Force, 1997, p. 23.)

The goal is to improve operational support of the warfighting CINCs, including the efficiency of weapon system support, by pursuing "best value" processes and products. Doing this calls for a full transition from deployed maintenance and "push" resupply to a method based

---

[15]In addition to specialized means, the Air Force employs more generalized means, such as precision-guided high-explosive weapons, that are already considered in the global attack planning area.

[16]Unclassified excerpts, DoD Directive S-3600.1.

on "accurate information, responsive production and daily, time-definite airlift."[17]

Achieving these goals for combat support involves

- **Supporting deployed forces**—Through time-definite resupply extending from the depot system to in-theater delivery. When a combat commander needs an item, "the system will reach back to the continental United States and deliver it where and when it is needed." (Department of the Air Force, 1997, p. 16).
- **Providing infrastructure to support air and space operations**—Through systems and logistics centers that support air and space operations.
- **Protecting forces**—Measures must be taken to defend air forces at home and abroad from conventional and unconventional attack. Combat support provides "protection for operational forces, means, and noncombatants." (CJCS, p. 2-132.) Similarly, it is tasked to prepare operationally significant defenses (CJCS, p. 2-132); provide security for operational forces and means (CJCS, p. 2-137); and protect and secure operationally critical installations, facilities, and systems (CJCS, p. 2-137).

Shape International Behavior

The Air Force contributes to the unified commands' shaping mission, helping to shape behavior by interacting with foreign militaries and by equipping foreign forces. Forward presence, which may also shape behavior, is subsumed under the planning area "attack anywhere on the globe."

Secretary Cohen has noted the continuing "great" need for forces able to

> shape the international environment in favorable ways, particularly in regions critical to U.S. interests. . . . In an increasingly interdependent world, U.S. forces must sustain credible military presence

---

[17]Briefing, "CORONA Fall '96 Long-Range Planning, Summary Themes," slide 29.

> in several critical regions in order to shape the international security environment in favorable ways. (Cohen, 1997, pp. 5–6.)

As General Shalikashvili has stated,

> our permanently stationed overseas forces, infrastructure and equipment, temporarily deployed forces, and the interactions between US and foreign militaries together demonstrate our commitments, strengthen our military capabilities, and enhance the organization of coalitions and multinational organizations to deter or defeat aggression. (Shalikashvili, 1995, p. 3.)

The shaping function has two principal elements—interaction with foreign militaries and equipping of foreign forces.

> Through training programs, multinational exercises, military-to-military contacts, defense attaché offices, and security assistance programs that include judicious foreign military sales, the United States can strengthen the self-defense capabilities of its friends and allies and increase its access and influence in a region. (Cohen, 1997, p. 6.)

BIBLIOGRAPHY

Air Mobility Command (AMC), *Air Mobility Master Plan,* Scott Air Force Base, Illinois, 1997.

Air Force Secretariat, "Building the 21st Century Air Force: Integrating AF Acquisition/Science and Technology and Strategic Planning," briefing, Washington, D.C., 1997.

Chairman of the Joint Chiefs Staff (CJCS), *Universal Joint Task List,* Manual 3500.04A, Version 3.0, Washington, D.C., 1996.

Clinton, William J., *A National Security Strategy for a New Century,* The White House, Washington, D.C., May 1997.

Cohen, William, *Annual Report to the President and Congress,* Washington, D.C., April 1997.

USAF Briefing, "CORONA Fall '96 Long-Range Planning, Summary Themes."

"CSAF Goals for 1997," *Policy Letter Digest,* January 1997.

Department of the Air Force, *Global Engagement: A Vision for the 21st Century Air Force,* Washington, D.C., 1997.

DoD Directive S-3600.1, *Information Operations (IO),* 3 December 1996.

Fogleman, Ronald R., and Sheila E. Widnall, "Cornerstones of Information Warfare," Department of the Air Force, 1997.

Headquarters (HQ) USAF/XOOR (Exercises), *Air Force Task List*, 6 November 1996.

Kent, Glenn A., and William E. Simmons, *Concepts of Operations: A More Coherent Framework for Defense Planning*, RAND, N-2026-AF, 1983.

Kent, Glenn A., and William E. Simmons, *A Framework for Enhancing Operational Capabilities*, RAND, R-4043-AF, 1991.

Lewis, Leslie, *Strategy-to-Tasks, A Methodology for Resource Allocation and Management*, RAND, P-7839, 1983.

Lewis, Leslie, J. A. Coggin, and C. R. Roll, *The United States Special Operations Command Resource Management Process, An Application of the Strategy-to-Tasks Framework*, RAND, MR-445-A/SOCOM, 1994.

Lewis, Leslie, Zalmay Khalilzad, and C. Robert Roll, *New Concept Development: A Planning Approach for the 21st Century Air Force*, RAND, MR-815-AF, 1997.

Lewis, Leslie, J. Schrader, J. A. Winnefeld, R. L. Kugler, and W. Fedorochko, *Analytic Architecture for Joint Staff Decision Support*, RAND, MR-511-JS, 1995.

Murdock, Clark, "Strategic Planning in the Air Force: An Update" (briefing), Washington, D.C., April 28, 1998.

"1996 Stakeholder's Report," *Guardian*, February 1997.

Pirnie, Bruce R., *An Objectives-Based Approach to Military Campaign Analysis*, MR-656-JS, 1996.

Shalikashvili, Gen. John M., *Joint Vision 2010*, Washington, D.C., July 1995.

Shalikashvili, Gen. John M., *National Military Strategy: Shape, Respond, Prepare Now: A Military Strategy for a New Era*, Washington, D.C., September 1997.

Special Operations Command, *SOF Vision 2020*, MacDill Air Force Base, Florida, 1997.

Thaler, David E., and David A. Shlapak, *Perspectives on Theater Air Campaign Planning*, RAND, MR-515-AF, 1995.

U.S. Space Command, *Vision for 2020*, Peterson Air Force Base, Colorado, 1997.

Widnall, Sheila E., speech given at Goodfellow Air Force Base, Texas, June 18, 1997.